Member Name:

_____

For Candee Cole: A true storyteller — S.W.

To Momchil: Thank you for inspiring me! — Y.P.

Scavenger Scout: Rock Hound

Text Copyright © 2018 by Shelby Wilde

Illustrations Copyright © 2018 by Yana Popova of YaPpy Arts

Book Design: Big Ape Design

First Edition

All rights reserved. Printed in China.

No part of this book may be used or reproduced in any manner whatsoever without written permission.

For more information, please visit us at www.shelbywildebooks.com.

Please direct inquiries to shelbywildeauthor@gmail.com.

ISBN: 978-1-7325168-0-9

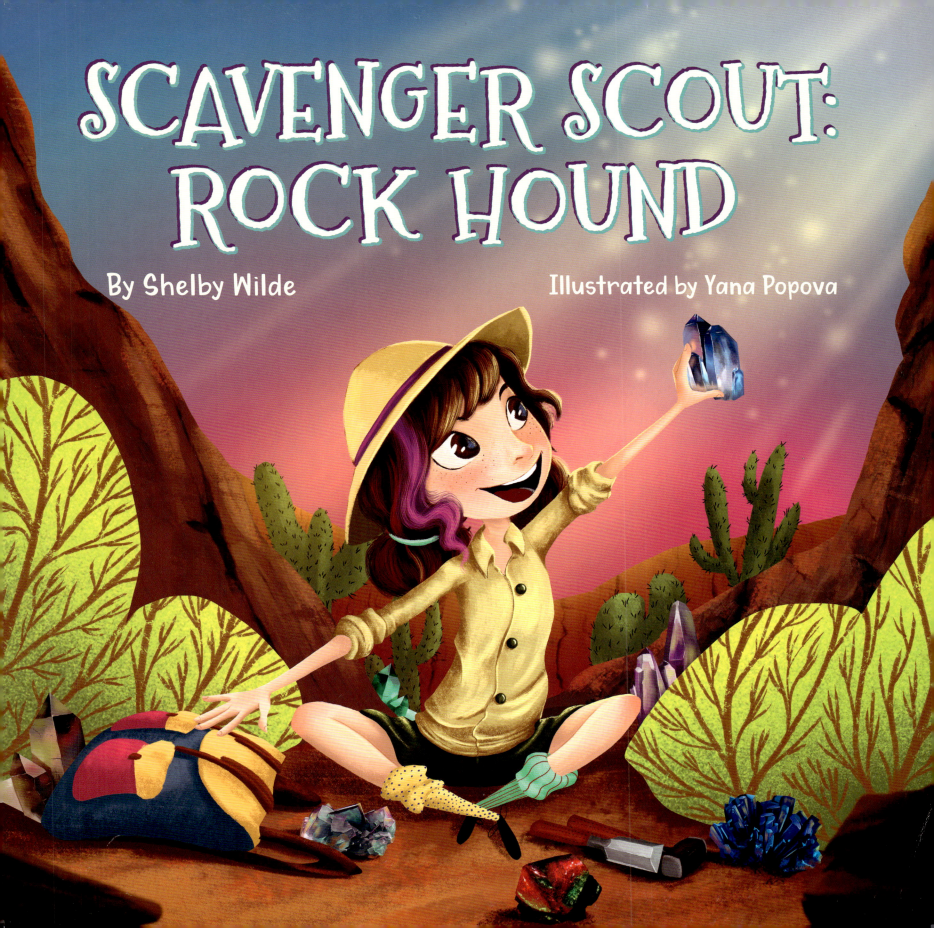

Do you **search** for minerals everywhere?

Do you own more rocks than underwear?

Do you 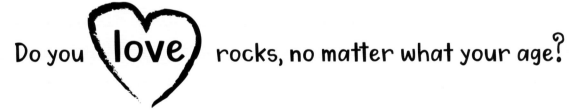 rocks, no matter what your age?

Well, follow Scout: find the on every page!

It all started with one **rock**.

One orange sparkly rock that I dug up in my back yard.

<u>I was **hooked**.</u>

You can call me **Scout**.

I'm a rock hound, mineral master, pebble pursuer.

Each rock tells a story, and I love to show and tell.

Where did I get all of these rocks?

Well, I'll tell you: I **hunted** down each one...

Azurite

Look for me ---→

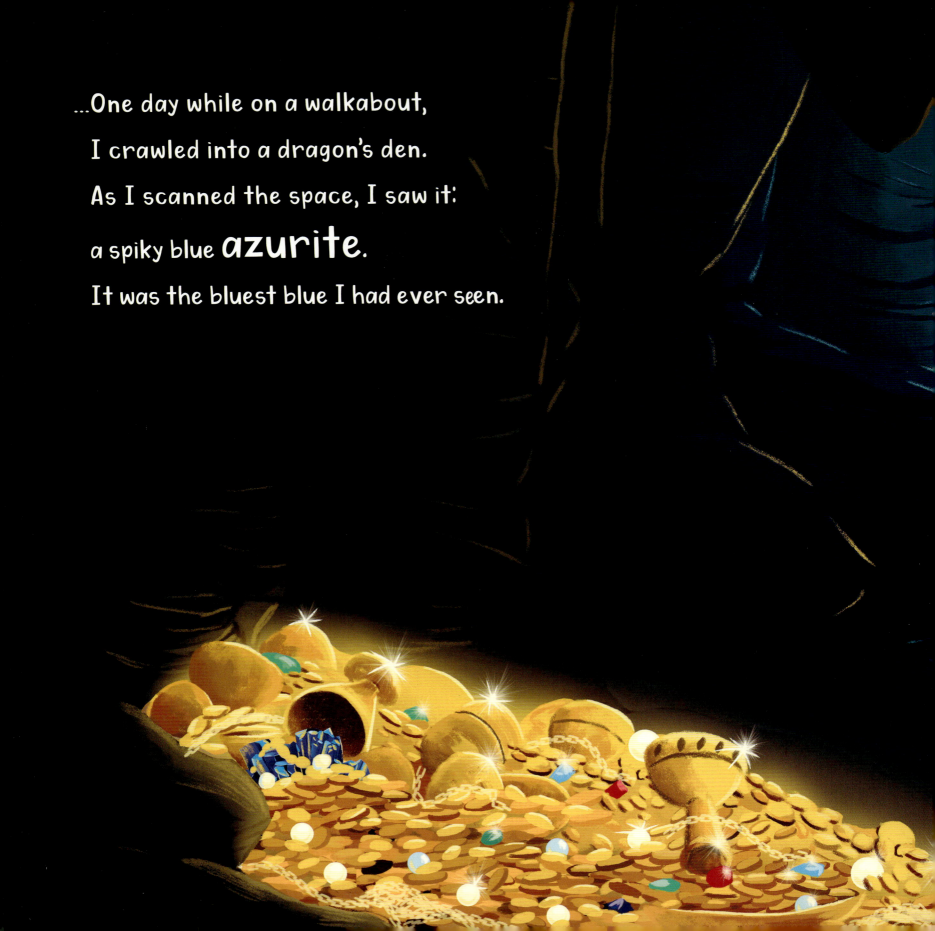

...One day while on a walkabout,

I crawled into a dragon's den.

As I scanned the space, I saw it:

a spiky blue **azurite**.

It was the bluest blue I had ever seen.

Stumbling over jeweled goblets, I crept closer.

I froze at a slithering sound —

a watchful dragon was leaving his gilded perch.

The azurite was the size of my baby brother (and probably just as much trouble). Fearing the dragon would change his mind, I quickly pulled out my crack hammer and hand chisel.

I fashioned a pulley with a counterweight and carefully lowered the stone into my pack.

From a deep, dark cave

to the ocean floor...

Fluorite

Look for me
----→

Under the sea, I secretly swam with mermaids and fishy friends. Hovering just above the ocean floor, my skin glowed green and purple — the colors of **fluorite**!

The giant crystal was stuck under an open clamshell studded with a pearl. I eagerly pulled out my pry bar.

The mermaids cried, "Those pearls belong to the sea. Please don't take them!"

"Gee, those pearls are swell, but I'm after my favorite: Fluorite."

The mermaids weren't convinced that their precious pearls were safe so I knew what I had to do: a rock swap. I grabbed a piece of coral and used my pry bar to swap it for the fluorite.

I hitched the fluorite to an empty ballast tank and raised the rock to the surface!

From the secret sea to the farthest reaches of space...

Alexandrite

Look for me ---→

I was moon-bound, on a mineral mission, when I spied a stone that looked like **alexandrite**.

Shining green one minute and flickering red the next. One of many that formed Saturn's rings, it floated with its friends — so tempting.

An alien watched me make a loop as I put together a plan. His voice sounded next to me, though he was still in his ship: "I have read your thoughts and must disagree, Saturn does not belong to you, but to the galaxy!"

I mashed the autopilot button and grabbed the cargo net from the back. It took three tries, but I snagged that stone!

From silent space
   to a collector's canyon...

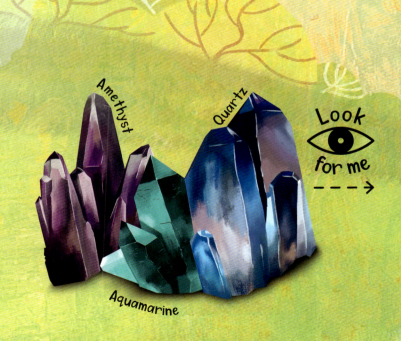

Amethyst
Aquamarine
Quartz
Look for me ----→

Heading west, I clutched a map that showed the way to a cluster of crystals. The sun was hot and X marked the spot where the clay and the river flowed.

The light glinted off a canyon wall in the distance. I leaned forward in the saddle, straining to see.

What greeted me around the bend was a dazzling sight:
A wall of crystal prisms, reflecting lots of light!

I picked up my pickaxe, but...
how was I going to get all of the crystals home?
Hey you, rock hound! Which crystal is YOUR
favorite and how would you get it home?

Rocks

Azurite

Color: Blue to dark blue

Hardness: 3.5 — 4

Where can you find it?
Australia, the Congo, Morocco, Namibia, United States (Arizona, New Mexico and Utah)

Fun fact: Azurite contains copper, which is what gives it the deep blue color.

Fluorite

Color: It is best known for the purple/green color combination, but it can be any color of the rainbow.

Hardness: 4

Where can you find it?
Argentina, Austria, Canada, China, England, France, Germany, Mexico, Morocco, Myanmar (Burma), Namibia, Russia, Spain, Switzerland, United States (Arizona, Colorado and New Mexico)

Fun fact: Fluorite glows when it is exposed to UV light.

Alexandrite

Color: Blue, Red, Green, Yellow, Pink, Purple, Gray, Multicolored

Hardness: 8.5

Where can you find it?
Brazil, India, Madagascar, Myanmar (Burma), Sri Lanka, Tanzania, Zimbabwe

Fun fact: Alexandrite can change color: it appears green/blue in daylight and red in incandescent light.

Mohs Hardness Scale (1-10)

This scale determines how hard a mineral is based on what rocks can scratch it. The higher the number, the harder the rock. Diamonds are a 10.

Tools

A) **Crack hammer**: These heavy hammers are used for breaking large rocks into smaller pieces. They are also used to drive chisels into rocks.

B) **Chisel**: Chisels are used for trimming, splitting and breaking rocks.

C) **Pointed-Tip Rock Hammer**: The flat end of the hammer is used for breaking rocks and for light chisel work. The pointed end is used to loosen rocks embedded in soil and for digging.

These tools should only be used by adults, while wearing appropriate safety gear.

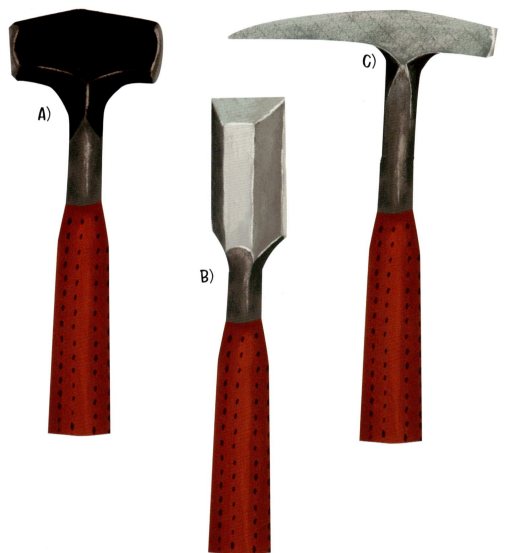

I'd like to thank the real Scout's family, both near and far, for their invaluable support. You loved this book before it was even written and your love shines through the pages.